535 Gaertner, Meg.
GAE Mirrors
      (Science all around)

DATE DUE			

MIRRORS

by Meg Gaertner

Cody Koala

An Imprint of Pop!
popbooksonline.com

abdobooks.com

Published by Pop!, a division of ABDO, PO Box 398166, Minneapolis, Minnesota 55439. Copyright © 2020 by POP, LLC. International copyrights reserved in all countries. No part of this book may be reproduced in any form without written permission from the publisher. Pop!™ is a trademark and logo of POP, LLC.

Printed in the United States of America, North Mankato, Minnesota

052019
092019

♻ THIS BOOK CONTAINS RECYCLED MATERIALS

Cover Photo: iStockphoto
Interior Photos: iStockphoto, 1, 5 (top), 5 (bottom left), 5 (bottom right), 6, 9, 10, 18; Red Line Editorial, 11, 13; Shutterstock Images, 14–15, 17, 20

Editor: Connor Stratton
Series Designer: Sarah Taplin

Library of Congress Control Number: 2018964777

Publisher's Cataloging-in-Publication Data

Names: Gaertner, Meg, author.
Title: Mirrors / by Meg Gaertner.
Description: Minneapolis, Minnesota : Pop!, 2020 | Series: Science all around | Includes online resources and index.
Identifiers: ISBN 9781532163593 (lib. bdg.) | ISBN 9781532165030 (ebook)
Subjects: LCSH: Mirrors--Juvenile literature. | Reflections--Juvenile literature. | Science--Juvenile literature.
Classification: DDC 535.32--dc23

Hello! My name is

Cody Koala

Pop open this book and you'll find QR codes like this one, loaded with information, so you can learn even more!

Scan this code* and others like it while you read, or visit the website below to make this book pop.

popbooksonline.com/mirrors

*Scanning QR codes requires a web-enabled smart device with a QR code reader app and a camera.

Table of Contents

Chapter 1

What Are Mirrors?

A mirror is a sheet of glass with a **backing**. The backing is a thin sheet of metal. The metal is often silver or aluminum. This backing **reflects** almost all light.

Watch a video here!

A mirror shows an image of any object in front of it. The image looks almost the same as the real object. But there is one big difference. The object's left side looks like its right side in the mirror.

Chapter 2

Light

Light moves in a straight line.

It hits the glass of a mirror.

Glass is usually **transparent**.

Transparent objects let light

pass through them.

Light moves fast. It could travel around Earth more than seven times in one second.

Complete an activity here!

Some objects are dark and **opaque**. They **absorb** most light. Other objects are shiny. They **reflect** most light.

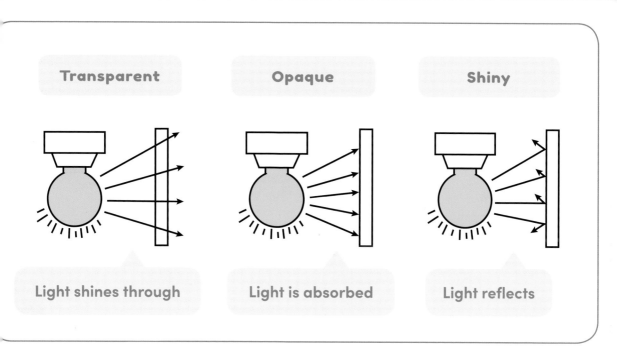

Transparent — Light shines through

Opaque — Light is absorbed

Shiny — Light reflects

Light bounces off shiny objects and reaches our eyes. Then we see reflections of things around the object.

Reflection

Objects **reflect** light in two different ways. Rough objects **scatter** light. The light bounces off the objects in all directions.

Learn more here!

Smooth, mirrorlike objects reflect light differently.

Light bounces straight back

in the direction it came from.

Using a Mirror

A girl stands in front of a mirror. Light shines from the sun outside. The light hits her face and is **reflected**.

Learn more here!

Some of the light bounces toward the mirror. The light passes through the mirror's **transparent** glass.

The light hits the mirror's metal **backing**. The metal backing reflects the light.

The light bounces straight back at the girl. The light hits her eyes. She sees herself in the mirror.

People have been making mirrors for thousands of years.

Making Connections

Text-to-Self

Do you ever use a mirror? What do you use it for?

Text-to-Text

Have you read other books about mirrors or light? What did you learn?

Text-to-World

Think of one job or activity people use mirrors to do. How would it be different if mirrors did not exist?

Glossary

absorb – to take in or soak up.

backing – a layer of material that forms the back of something.

opaque – not letting any light pass through.

reflect – to bounce back without absorbing.

scatter – to separate in all directions.

transparent – letting all light pass through.

Index

Online Resources

popbooksonline.com

Thanks for reading this Cody Koala book!

Scan this code* and others like it in this book, or visit the website below to make this book pop!

popbooksonline.com/mirrors

*Scanning QR codes requires a web-enabled smart device with a QR code reader app and a camera.